THE URGE TO DISPERSE

By

Sheila Newman,

Environmental Sociologist

Incest avoidance and the Westermarck effect as indicators of biological algorithms for population dispersal, spacing and size, in many species including humans

By the same author

Sheila Newman, Ed., *The Final Energy Crisis,* 2nd Edition, Pluto Press, UK, 2008

Andrew McKillop with Sheila Newman, Eds., *The Final Energy Crisis,* Pluto Press, UK, 2005

The Growth Lobby and its Absence in Australia and France, Environmental Sociology thesis, Swinburne University, 2002, http://adt.lib.swin.edu.au/public/adt-VSWT20060710.144805/index.html

ISBN 978-1-4467-8413-6

Candobetter press, http://candobetter.net

22 March 2011

Preface

This book looks at the role of incest avoidance and the Westermarck effect as evidence of biological algorithms[1] for population spacing and dispersal patterns which, if undisturbed, would keep populations naturally within carrying capacity. It also explains why populations may overshoot or fail to establish. The implications are, broadly, for maintenance of steady state populations – human and others -within a rich and diverse ecology. The book uses both social and biological science.

(The author would welcome collaboration in expressing and testing the theory using computerized population modelling).

TITLE PAGE

1. Review of the Literature (Social and Zoological)

Social Theory:

All societies have laws relating to incest, overwhelmingly banning it as 'taboo', with some exceptions, usually in ruling castes, famously with the Egyptian pharaohs, where it was actually prescribed.

Levi-Strauss: Incest Prohibition as the Socio-sexual Organiser

The theory of incest avoidance as a major ordering force of population spacing is a well-known anthropological one which has been written about by, among others, Levi-Strauss. The theory is that the sexual and the social intersect dynamically. Strauss identified the prohibition of incest as the fundamental socio-sexual 'organiser'. Prohibition of incest – that is to say, of sexual relations with blood-relatives or with the same 'class' (e.g. relatives by marriage) – acts universally as the point of encounter where the sexual meets the social. In *Elementary structures of families*, (1949) (*Les Structures élémentaires de la parenté*, 1949), Claude Levi-Strauss interpreted this as the decisive moment of the passing of nature into culture, and as the pivot of a movement that required the exchange of women. From this point, he argued, stems the organised system of marriage based alliances and inheritance – that is, the social field. A British ethnologue, Robin Fox, treated the problem from an evolutionary perspective: Groups with very strong drives would have established moderating mechanisms, such as the prohibition of incest, which would have led to their 'adaptation'. (*Anthropologie de la parenté*, 1967). [2]

The explanation above assumes that incest avoidance was the result of prohibitions, but, as I will elaborate below, zoological observation and testing of hypotheses indicates that incest avoidance is present in other species, where conscious prohibition is not used as an explanation. Tests reveal suppression of ovulation and testosterone production in the presence of close relatives in non-human species.[3] This is a reason why I prefer the term, "incest avoidance" over the term "incest prohibition".

Biological theory:

What is the biological function of the almost universal taboo, also known as 'incest avoidance'?

Most people think that they know, and that statement includes psychiatrists, zoologists and sociologists. The usual explanation is that the incest taboo is a social response to the costs of incest. It is inferred that societies observed that the outcome of incestuous relations was unhealthy children. Fault may be found with this explanation on several accounts, but the most important problems with it are that many creatures practice incest avoidance as assiduously as humans, including such tiny and presumably unscientific organisms, as cockroaches, which we would not expect to make conscious decisions on the grounds of observation. But, perhaps we underestimate cockroaches.

The social learning theory of incest taboos is in any case challenged by the theory that incest is more likely to breed-in healthy chromosomes and to breed-out defects than to compound ever more complex defects. This would be because, where genetically healthy relatives interbreed, they will not have unhealthy offspring. Where relatives carrying identical but recessive unhealthy chromosomes interbreed, their offspring are likely to be unhealthy with reduced survival and reproductive chances. The reduced reproductive chances could result in the breeding out of the problematic genes from the gene pool. [4]

Against the **culturally mediated genetic diversity impulse explanation** is the application of incest taboos to 'in-laws', in a more or less symmetrical fashion to true incest avoidance. By breeding with in-laws (who are not blood related) the gene-pool would increase, but this opportunity is avoided and further dispersal remains necessary to find mates. In fact in-laws are in a sense indirectly blood-related by marriage, and potential producers of directly blood-related offspring.

Anthropologist Levi-Strauss rejected biological explanations of human social behaviour. Unaware of the occurrence of incest avoidance in so-called primitive societies and in other species, he believed that incest avoidance was a cultural evolution from savagery to civilized society. [5] Marx also theorized this. Freud believed that culture had invented the incest taboo in order to mitigate against an extremely strong drive to incest in humans. Levi-Strauss believed that humans attenuated hostility between groups and encouraged alliances by having their children marry outside their own group. He also thought that it would become too complex and difficult to identify the status of a group's members if they were all very closely interrelated. He thought the outcome might be for the family unit to break down under the stress of sexual and territorial conflicts. Those explanations are still reasonably useful to explain an outcome of incest avoidance, which is social organization.[6] The seeking of mates within one's own group is called 'endogamy'. Dispersal to find mates outside one's own group is called 'exogamy'.

Interestingly, Freud was inspired to his theory of incestuous impulse (*oedipal conflict*) by his interpretation of the story of *Oedipus Rex* in Greek mythology. Ironically, the Greeks actually got the mechanism right, for Oedipus was only able to be attracted to his mother because he was not raised by her. (See further on for the Westermarck effect.) Freud's theory had only retained the fact of an incestuous attraction. Freud thence theorized the presence of an 'oedipal complex' with its feminine the 'electra complex' as a part of human development in order to explain the frequency of reports of incest from his patients, which he decided were most probably fantasies.

Freud was unaware of the Westermarck effect, (see below) however this effect might have provided him with another explanation for the rate of incest reports he experienced in late 19th century Vienna. In an era and social strata, in households where nurses took over the mother's role and fathers often spent much of their time away from home at work, or otherwise socially distanced from their children, the Westermarck effect may not have been effective. If that was so, then the barriers to incest between father and daughter, mother and son, might have been reduced. In modern industrial society, the pattern where both parents work long hours away from home and children, may carry a similar danger.

The Westermarck effect

Towards the end of the 19th century, Finnish sociologist, Edvard Westermarck, whilst conducting research into endogamy and exogamy, discovered a phenomenon which came to be called the *Westermarck effect*.

He was able to show that incest avoidance applied to people raised together, whether or not they were genetically related. The effect has since been observed many times over, notably in a longitudinal study of kibbutzes by Shepher in 1983.[7] In this study of children raised communally in peer groups, of the 3,000 marriages of those children in later life, only 14 were to members of the person's childhood kibbutz peer-group, and none of those 14 had been reared together during the first six years of life.

The absence of the Westermarck effect coincides with the absence of incest avoidance responses where close blood relatives are raised apart from each other.

What does this mean? Does it mean that nature is easily fooled about blood relationships and causes meaningless oppositions but allows dangerous attractions?

If we look at incest avoidance and the Westermarck effect more broadly - as impulses for population dispersal rather than moral or social arbiters - then the mirror phenomenon of blood-relation avoidance, the adoptive-relation avoidance, and the indirect blood-relation avoidance (in-law avoidance) seem to more consistent and coherent. I don't think that we can, however, diminish the structural importance of incest-avoidance, since this seems clearly to set the initial pattern of population dispersal in time and space.

Although mainstream sociology has never really embraced the theory of a biological basis to incest avoidance, research has progressed in zoology, leaving Levi-Strauss, Freud and Marx behind, but not Westermarck.

2. Incest avoidance and the Westermarck effect as a population dispersal mechanism affecting population size and political organization through inheritance of territorial rights

This book looks at research and theory into the incidence and impact on social structure of incest avoidance, the Westermarck effect, and inheritance patterns. It advances a new theory of incest avoidance and the Westermarck effect in population dispersal as an independent variable of land-tenure, pattern of land-settlement, population size and density. It hypothesises that the way human societies are organised depends greatly on how they link land-use to inheritance, which in turn links the land to the social structure of the family, clan, tribe and village. Also in this book are new suggestions about the effect that disruption of incest avoidance/Westermarck effect and their relation to place could be having on modern populations in aggregate and on individuals. The book does not focus on incest itself and especially not in the social sense of a criminal or depraved act. This is not a psychiatric or a criminological investigation. It is also not a discussion about the risks of inherited disease or deformity due to inbreeding. **Essentially it is about how the need to go away from the primary family unit to find a sexual partner is both a limiter of reproductive opportunity and a primary impulse for population dispersal.**

This theory can be used to reinterpret early, medieval and modern history with particular reference to theories of societal collapse and the development of expansive and conservative economies and populations.[8]

Identifying a New Function of Incest-Avoidance and the Westermarck effect: Population Dispersal

To state what may seem obvious and trivial: incest avoidance would be almost automatically guaranteed in large populations due to the laws of probability.

Conversely, in small, isolated populations incest is very likely without incest avoidance strategies and unavoidable under normal conditions of fertility for communities under a certain size to avoid extinction. In such cases exceptions to the incest avoidance rule may or may not occur. Where these exceptions occur they cause caste systems in humans, sub-species in other animals, initial territorial aggregation and finally overshoot. (See further on for examples under the heading "Isolated Populations".)

In this book I want to offer material to support the idea that incest avoidance and the Westermarck effect may be part of a response to genetic algorithms which underpin the organisation of human settlement; i.e. that incest avoidance is an instinctive norm in humans.

Incest avoidance and the Westermarck effect and the related population dispersal seems to be a dynamic deeply embedded in most or all sexually reproducing species, even including the reluctance of hermaphrodite plants to 'self' themselves.

The impulsion for population dispersal saves species from isolation and consequent annihilation through local accidents. It also fits the 'selfish gene' theory of Dawkins, by promoting maximum survival opportunity for genetic material in a multitude of life-forms in many different places. It is also logical and practical, or coral chains and human settlements would (oxygen, food, and musculo-skeletal structures permitting) reach up vertically to the moon and beyond.

Numerous studies, some of which are discussed in here, demonstrate that incest avoidance occurs in many species. Examples discussed here include the acorn woodpecker, the superb fairy wren, voles, mice, butterflies, marmosets and primates and could have included toads, cockroaches and more.[9] For instance, in hermaphrodite species, including plants:

> "Another type of inbreeding avoidance is avoidance of selfing in hermaphrodite species. This is of great interest in particular because low selfing rates or even complete selfing avoidance are very commonly found in plants, and also in most hermaphrodite animal taxa (reviewed in Jarne & Charlesworth 1993)[10]. Avoidance of selfing can be expected to have the same effect in

our model as sibmating avoidance in dioecious species. This is because from the cytoplasm's viewpoint, a hermaphrodite is essentially the same as a pair of siblings in the context of reproduction. Moreover, at least one of the mechanisms of selfing avoidance—self-incompatibility—can be expected to also lead to incompatibility between closely related individuals, thus resulting in even higher outbreeding rates than with mere selfing avoidance."[11]

Human land-tenure systems as outgrowths of biological algorithms

Some of the research published shows clearly that incest avoidance impacts on population spacing (i.e. density and fertility), even though the research has been done for different reasons – often to elucidate scientific or political arguments about whether incest avoidance is genetically or culturally based.

Although my interest is not in the political polemics of nature vs nurture, the politics of the nature versus nurture argument regarding the origins of incest avoidance invite consideration of the role of culture in land-tenure patterns and of biology in culture. **How densely we occupy land is a function of population spacing. The pattern in which populations disperse and organize in space is initiated by incest avoidance and the Westermarck effect.**

Thus our human land-tenure systems seem to be outgrowths of biologically based algorithms. Work done on the socio-sexual organization of birds, monkeys, apes and other animals, gives reason to suppose that a system of hormonal feedback from the environment probably informs the application of genetic algorithms to local conditions.

Population Numbers and Regulation

Some of this work has been done to evolve theories as to why some populations overshoot their resource base and others do not.

One such theory is the 'mechanistic' predator-prey theory, which hypothesises that some or many populations may be kept in check simply by being slain and eaten by predators. Dependence of the predators on the predated population in turn keeps the predators in check. In its crudest expression this theory ignores in-migration, out-migration, and fails to establish whether the initial population arose naturally or chaotically, which impacts on rate of incest avoidance and the Westermarck effect, often supposing that all populations simply breed as much as they can, if they do not benefit from modern contraception, albeit with seasonal variations.[12]

A more sophisticated appraisal of what may be involved in population regulation is to be found in Pimentel, "Population Regulation and Genetic Feedback".[13] The author identifies a number of rules. One is that most species are quite rare, relatively or 'by whatever criterion they are judged'.[14] This rule helps to construct the idea that huge numbers involved in overshoot by a species are probably rare and do not last for long. Another is that nearly all animals feed off live material. This observation is important because dead material cannot evolve genetically in response to predation. Pimentel describes field observations and laboratory tests which show that predated populations evolve in response to a particular predator "only if the numbers of the animal are sufficient to exert some selective pressure on the host." Using a variety of examples, he observes that the dominant control mechanism operating initially is "competition" (meaning selection), "but genetic feedback became dominant with time and through evolution." He observes that "subtle genetic changes" affect the predator, and gives this example:

> "For instance, when young pea aphids (*Acyrthosiphum pisum*) were placed on a common crop variety of alfalfa (*Medicago sativa*), they produced a mean of 290 offspring in 10 days, whereas the same number of aphids for a similar period on a resistant alfalfa variety produced a mean of only two offspring. In another example, the mean rate of oviposition (eggs per generation) of the chinch bug (*Blissus leucopterus*) on a susceptible strain of sorghum (*Sorghum vulgare*) was about 100, whereas on a resistant strain the mean oviposition was less than one. In both, reproduction in the animals feeding on the resistant plant hosts decreased more than 99 per cent. This reduced reproduction obviously would have dramatic effects on the population dynamics of the feeding animal populations."[15]

In Hopfenberg and Pimentel, "Human Population Numbers as a Function of Food Supply,"[16] the authors cite evidence of other species decreasing their fertility by a variety of means which would suggest a hormonal feedback mechanism from food availability in the environment.

> "Some species self-regulate their number to their food resources by maintaining home ranges. Chitty (1995) reported that excess young voles, for example, are forced to leave the home range of their parents... . Possibly more germane is the evidence that a sudden improvement of diet in sheep causes an increased ovulation rate (Schinkel, 1963) and that fasting in mice for relatively short periods of time prior to mating resulted in depression of male libido and reduced conception in females (Christian et al., 1965)." [17]

They attest that this simple relationship between food supply and fertility is also seen in hunter-gatherers who do not have massive infrastructure to complicate their interactions with the environment.

> "The populations of human cultures described as hunter-gatherers were limited to the food resources available (Lee, 1969; Lee and DeVore, 1976; Pimentel and Pimentel, 1996). Where these cultures still exist untouched, this continues to hold true." [18]

Biologists know that population density itself is a major indication of soil and climate fertility, that is, of food availability.[19] Density is also a determinant of growth rate in naturally organized populations, by which I mean those unaffected by artificial or highly disorganizing factors.[20]

> "The negative relationship between population density and population growth rate is at the heart of population biology." (Hone J, Sibly RM., 2002) [21]

Need to consider a different kind of mechanism

15

These studies and observations, however, do not identify a mechanism which would limit fertility in order to avoid indefinitely the experience of restricted calories. If food shortage were required before fertility dropped then clans and herds would be permanently on the verge of starvation. That kind of stress would not be conducive to genetic survival.

> "... Iwamoto (1978) has shown that monkey troop size increases rapidly after artificial provisioning, but the level of consumption efficiency of the troop is always maintained lower than the critical point in both the artificial and natural habitat. Starvation within the troop simply does not occur if the rate of food availability is held relatively constant." [22]

There has to be another mechanism whereby species can adjust collectively to local environmental constraints of their ranges without constantly risking starvation.[23]

As early as 1940, in "Australia: Ecology, spacing mechanisms and adaptive behaviour in aboriginal land tenure", Joseph B. Birdsell's meticulous study of marriage laws and patterns in Australian aborigines over a wide range of climates showed that strictness of incest avoidance strongly corresponded with environmental fertility as measured by rainfall.[24] Naturally soil quality and quantity and temperature were other quotients but rainfall was the indicator that Birdsell measured. We can see that rainfall would be simpler to quantify for such research and possibly a more reliably distributed indicator than soil or temperature range. In these regions it is actually the decisive factor in environmental fertility.

Which are likely indicators of environmental fertility that may mediate fertility and space in humans and other species?

From Birdsell's observations above, it seems that rainfall [plus soil] is probably an indicator of environmental fertility which could be objectively measured in food availability and that potential, average and seasonal food availability restricts animal fertility in some way well before the point of food shortage. Birdsell also identifies (in humans) the fact that the social mechanism (also present in other animals) of incest avoidance limits fertility opportunities and can be varied in strictness in response to conditions.

What might be a mechanism whereby incest avoidance adjusts to environmental variations over regions?

It does not seem likely that the many species of animals which practice incest avoidance make complex conscious calculations about carrying capacity and then respond by devising such codes of conduct. More probable is that hormones are a mediating factor between reliability of food availability and fertility. They might be thought of as a chemical messenger.

Living creatures are biochemical systems which are regulated by hormones. Not all hormones are concerned with sexual reproduction, of course. Hormones regulate absolutely all our functions, from appetite, sugar uptake and digestion, to sweating, tissue growth, body shape, muscle quality, and sexual maturity and sexual behaviour. Hormones are responsive to external and internal environmental conditions. Some hormonal responses are obvious and well known, such as the impact that spring has on reproduction in most living things – plant and animal – and the response of sexual attraction that occurs on encountering a prospective mate.

Studies in animals and in people show that hormones are also affected by the presence of close family. One of these effects seems to be the suppression of oestrus in incest avoidance and the Westermarck Effect. Relevant studies on acorn woodpeckers, marmosets and apes are discussed in this context further on.

Based on the role of hormones described above and the objective data from the animal studies below, my hypothesis or answer to the questions I have framed above is that hormones will deliver more or less fertility according to availability of living space. Space (territory) required per individual will be affected by density and reliability of food distribution, and all of this will be mediated by some degree of incest avoidance/Westermarck effect, which is also related to social dominance. (Other forms of dominance such as caste and gender may also operate.

Opportunities for meeting may also be limited by gender specific circuits and territories, which can themselves be seasonal.)

A minimum physical distance from blood relatives or 'in-laws' would be required before sexual activity becomes likely. If a subject who is of an age to reproduce is dominated (i.e. within the physical sway of a parent) ovulation and sexual maturity may actually be measurably chemically inhibited.[25]

In some species sexual maturity or ovulation is delayed or permanently postponed where not enough territory is available. Where ovulation does not occur in females, even though males may have viable testosterone levels, their sexual activity will be decreased or absent. This is presumably related to opportunity and opportunity is in part defined by the presence of oestrus (being in season) and oestrus is limited by the close presence of relatives.[26]

Studies of Incest avoidance in other species

Biologist and ecologist, E.O. Wilson reports in "Nature Matters" on the high incidence of incest avoidance algorithms in many species which have been detected in biological anthropology, sociobiology, cognitive psychology and neuroscience studies[27]:

> "These algorithms can be blocked or reversed only at the peril of mental health. An example is the negative imprinting that forms the basis of incest avoidance, as follows: When either of two persons lives in close domestic proximity during the first 30 months' life of either one, both are unable to form close sexual bonding later in their lives. The phenomenon, known as the Westermarck effect in honour of the Finnish anthropologist who discovered it a century ago, is evidently widespread if not universal in human beings. Equally impressive, it is shared by all other primate species whose sexual behavioural development has been closely studied.
>
> The nature of the automatic incest-avoidance process, as well as its evolutionary origins, now seems well established by solid research".

He concludes that there is a clear adaptive advantage "to those who react to it correctly". He believes this is the avoidance of "inbreeding, fewer homozygous defective genes, and more healthy children". He writes that the *"Westermarck effect is an example of an epigenetic rule, defined as an inherited regularity of development."*

Whilst I agree that incest avoidance promotes genetic diversity and thereby reduces the disease risks of inbreeding, as I have already mentioned, my own conclusion is that population spacing is the most important effect. Inbreeding does carry risks but population spacing is absolutely vital to social organisation.

Acorn Woodpeckers – incest avoidance in a complex social order

Walt Koenig and Joey Haydock, in the "Social Behavior of the Cooperatively Breeding Acorn Woodpecker," [28] give a riveting report on a study which has been going on since 1971. Their observations show that 'mere' birds maintain symmetrical incest avoidance/Westermark Effect in the most complex of social orders. The reader may draw the conclusion as I do that incest avoidance is the basis upon which this social order is built. The study shows that sexual maturity does not occur without the presence of suitors who are genetically sufficiently distant. This implies complex hormonal suppression. Specific hormone suppression is actually tested and measured in a study following this one.

"Acorn Woodpeckers are common residents of oak woodlands in western North America.

"Groups engage in a many communal activities, including territorial defence, feeding of young at the communal nest and acorn storage in special trees known as granaries. Stored acorns are an important food resource, both during the winter and for successful reproduction the following spring. Groups can even breed in the fall when the acorn crop is particularly good.

"They have one of the most bizarre mating systems of any bird in the world. They are co-operative breeders and live in groups composed of up to six co-operative breeder males, three joint-nesting females, and non-breeding helpers of both sexes.

"Co-breeder males are brothers and/or fathers and their sons competing for matings with the joint-nesting females, who are sisters or a mother and her daughter who lay their eggs in the same nest cavity. Offspring produced from this communal nest may remain in their natal group for several years as non-breeding helpers, during which time they help feed younger siblings at subsequent nests.

"This kind of mating system is known as polygyandry. All individuals within the group are close relatives except that co-breeder males are not related to joint-nesting females. **Incest avoidance is maintained because helpers only inherit and become co-breeders following reproductive vacancies when the breeders of the opposite sex die and are replaced by unrelated birds from elsewhere.** Reproductive vacancies are often filled by a unisexual set of siblings who compete against other sibling groups in spectacular events called power struggles. Winners of power struggles become co-breeders in the new group; losers return home and resume non-breeding helper status. (…)"

The mathematical and spatial consequences of these practices would be a reduction in fertility opportunities which would otherwise be available if any male could mate with any female. Within the prevailing arrangements more territory is available per breeding group than would be available if most birds bred.

As odd as the Acorn Cuckoo's households may seem to us, they may not be all that unusual among birds. The Superb Fairy Wren was recently the subject of a study which demonstrated very similar arrangements in Cockburn, Osmond, Mulder, Green and Double, "Divorce, dispersal and incest avoidance in the cooperatively breeding superb fairy-wren, Malurus cyaneus".[29]

I have quoted their useful numbered summary.

"1. Between 1988 and 2001, we studied social relationships in the superb fairy-wren *Malurus cyaneus* (Latham), a cooperative breeder with male helpers in which extra-group fertilizations are more common than within-pair fertilizations.

2. Unlike other fairy-wren species, females never bred on their natal territory. First-year females dispersed either directly from their natal territory to a breeding vacancy or to a foreign 'staging-post' territory where they spent their first winter as a subordinate. Females dispersing to a foreign territory settled in larger groups. Females on foreign territories inherited the territory if the dominant female died, and were sometimes able to split the territory into two by pairing with a helper male. However, most dispersed again to obtain a vacancy.

3. Females dispersing from a staging post usually gained a neighbouring vacancy, but females gaining a vacancy directly from their natal territory travelled further, perhaps to avoid pairing or mating with related males.

4. Females frequently divorced their partner, although the majority of relationships were terminated by the death of one of the pair. If death did not intervene, one-third of pairings were terminated by female-initiated divorce within 1000 days.

5. Three divorce syndromes were recognized. First, females that failed to obtain a preferred territory moved to territories with more helpers. Secondly, females that became paired to their sons when their partner died usually divorced away from them. Thirdly, females that have been in a long relationship divorce once a son has gained the senior helper position.

6. Dispersal to avoid pairing with sons is consistent with incest avoidance. However, there may be two additional benefits. Mothers do not mate with their sons, so dispersal by the mother

liberates her sons to compete for within-group matings. Further, divorcing once their son has become a breeder or a senior helper allows the female to start sons in a queue for dominance on another territory. Females that do not take this option face constraints on their ability to recruit more sons into the local neighbourhood."

This above should make it clear that the practice of females travelling to avoid incest would profoundly affect the spatial patterns of settlement and the rate of population growth.

The cooperative breeding might also improve the chances of fledglings reaching adulthood – and thence raise the population growth rate - although this is a moot point since, at least in the case of the Acorn Cuckoos, nesting female birds frequently got rid of their sisters' eggs from the nest.

Incest Avoidance in Non-human Primates

In "Constraints on control: factors influencing reproductive success in male mandrills (Mandrillus sphinx)"[30], Marie Charpentier, Patricia Peignot, Martine Hossaert-McKey, Olivier Gimenez, Joanna M. Setchell, and E. Jean Wickings, write that the *"Mechanisms of inbreeding avoidance are well documented in vertebrate societies"*, and cite a number of well established references.[31]

They inform us that the social structure of most primates, including macaques and the mandrills in this study, is such that the males tend to disperse after adolescence to distant groups, whereas the females remain mainly with the home group. [32] (The opposite occurs with chimpanzees and is more common in human dispersal.[33]) They confirm that in the mandrills they studied, social rank is inherited by females via a female line, but males generally leave the natal group to seek positions in other groups.

Their study showed that incest avoidance was marked where there was 50% relatedness (as in siblings or parents to children), and that incest avoidance decreased as the blood relatedness decreased. The subjects were observed in a colony that was artificially isolated, so that the male tradition of joining distant colonies could not be carried out. Tests showed that males who were obliged to remain with the colony also observed incest avoidance, which demonstrated a knowledge of kinship, according to the authors.

"Several studies have shown that maternal relatives avoid mating with one another (rhesus macaques: Smith, 1995; red colobus, Procolobus badius temminckii: Starin, 2001; Japanese macaques: Takahata et al., 2002; and see for review: Moore, 1993; van Noordwijk and van Schaik, 2004), but less is known concerning patterns of inbreeding avoidance between paternal relatives (but see Alberts, 1999). In this study, we showed that the probability of paternity by a dominant male decreased when he was related to the dam at $R = .5$ (the highest possible relatedness coefficient in our study). Smith (1995) showed in rhesus macaques that the intensity of inbreeding avoidance was directly correlated with the closeness of kinship, as in the mandrills studied here."

The authors also write:

"The incest avoidance shown here demonstrates that female mandrills may exercise an active choice of partner, avoiding mating with close relatives, as raising an inbred offspring may be costly. A detailed study of female mate choice is indicated to determine the effects of relatedness."

(Obviously I don't agree with the inference that females avoid incest in order to avoid genetic disease problems).

"Surprisingly, in this study, we also showed that the more closely the dominant male was related to adolescent males in the group, the higher was his probability of paternity. This is the opposite of the finding concerning adult males. In the mandrill colony, different mating strategies appear to be employed at different times during a subordinate male's lifespan. A hypothetical pattern could be proposed. During adolescence, males appear to support

23

a closely related dominant male and hence may avoid aggressive interactions that they cannot win, but on reaching adulthood, the situation is reversed and subordinate males may compete successfully for females, dominant males being less vigilant with closely related subordinates."

My comment is that this also shows kinship and territory relationships. It also resembles the strategy adopted by young human males towards their fathers. The statement that "the more closely the dominant male was related to adolescent males in the group, the higher was [the dominant male's] probability of paternity" is an indication to me of lower sexual activity in male offspring on the parental home range, except of course in the father or other dominant relative.

Marmosets and Hormonal Suppression in the presence of dominant close family

"Common marmosets live in social groups in which all group members help to raise the babies of a single dominant female. In marmoset groups only the dominant male and female breed whereas lower ranking or subordinate marmosets do not because their reproduction is repressed." [34]

In this report entitled "The Common Marmoset, Callithrix jacchus, Current Research",[35] sexual immaturity or infertility in the presence of dominant close family is actually linked to suppression of oestrus in females, with measurably smaller ovaries. The infertility in female marmosets is similarly affected by the presence of non-related dominant females. (Note that, in humans, girls raised in the presence of their blood related fathers reach menarche (i.e. menstruate for the first time) later than girls raised in the presence of a non-blood related father.[36])

Male marmosets have reduced rates of sexual activity in the presence of dominant close family where the chances of incest are high. Their testosterone levels, however, are not affected. This may indicate a behavioural avoidance of incest, but it could be due to the lack of available females in season, which seems itself to be a response to being unable to find enough territory to escape parental presence or the more dominant unrelated females. This behavioural choice supports the Marmoset family structure where only one couple breeds and the other relatives assist in the raising of that couple's babies.

Here is an extract from this interesting report.

"Reproductive Constraints:

"Subordinate or low ranking male and female marmosets do not reproduce, however, researchers believe it is because of several reasons. Subordinate males and females appear not to breed because of the risk of incest and behavioural limitations whereas subordinate females also have hormonal constraints. Common marmosets live in social groups in which all group members help to raise the babies of a single dominant female. In marmoset groups only the dominant male and female breed whereas lower ranking or subordinate marmosets do not because their reproduction is repressed. This suppression of reproduction is the case in many primate and non-primate mammals including yellow baboons. Dominant marmosets, unlike dominants in other species, however, do not appear to use harassment as a means to keep the subordinates from breeding.

"Researchers have shown that the ovaries of the subordinate female marmosets are about half the size of the dominant females. Blood samples taken from subordinates to detail hormone levels also revealed that subordinates do not ovulate. When the subordinate females are taken out of the presence of a dominant female and placed on their own they will ovulate. Researchers, therefore, believe that reproductive suppression in subordinate females is due to the hormones of the subordinate female. One hormone believed to be involved is released from the brain and stops the release of hormones stored in reproductive organs. When researchers gave large doses of this hormone to subordinate female marmosets they began to ovulate and when the levels of the hormone were reduced the female stopped ovulating.

"Researchers believe the lack of this particular hormone has something to do with the reproductive suppression of subordinate female marmosets.

"Researchers have not pinpointed which cues, visual, behavioral, or olfactory (smell), produced by dominant female marmosets cause subordinates to stop ovulating and, in turn, stop reproducing.

"Like the subordinate female common marmosets, subordinate males also do not reproduce.

"Researchers, therefore, wanted to know the causes of male reproductive suppression. Researchers thought that reproductive suppression of subordinate males was due to cues given by dominant males or as an incest avoidance behaviour to keep family members from mating. For the study subordinate males and their fathers were tested either alone or together in a cage and were joined by another female familiar to them (like their mother or mate) or an unrelated and non-familiar female. The number of sexual behaviours the males engaged in was recorded and blood samples were taken to measure hormone levels. Researchers found that the sons engaged in very low rates of sexual behaviour with familiar females whereas their fathers (the dominant male) engaged in higher levels. This might imply that the subordinate male does not mate with familiar females because of the chance of being related. When fathers and sons, however, were tested individually with unrelated females, both engaged in approximately the same number of sexual behaviours.

"This would suggest that subordinate male marmosets have no problem mating if it is with a non-familiar female. Finally, the hormone levels were examined and the researchers found no difference between the levels of fathers and sons. Researchers think that because subordinate male marmosets do not engage in much sexual behaviour with familiar females and because hormone levels are similar between dominant and subordinate males that subordinate males are reproductively suppressed to avoid incest.

"Researchers now have evidence to believe that subordinate male common marmosets do not reproduce so as to avoid incest. Subordinate females, however, are thought to be reproductively suppressed because of hormones. Presently, researchers are still studying the cause of reproductive suppression in male and female common marmosets."

These zoological reports demonstrate the presence of incest avoidance in several social species. The practice of incest avoidance reduces potential population size and density, providing more territory per animal than would otherwise be available. Without incest avoidance, populations would reach much higher density, reducing average territory per capita. Weight, size, quality of life and longevity would suffer and one would expect increased aggression.

We can also see political behaviours accompanying incest avoidance, such as submission or harassment (in baboons, for instance). In marmosets and cooperatively breeding birds, there is a politico-social adaptation. These organisational patterns seem to be algorithms with incest avoidance and the Westermarck effect an obvious denominator.

3. Towards a new social theory on population density and geometric patterning.

How densely we occupy land is a function of population spacing. The pattern in which populations disperse and organize in space is a function of incest avoidance. For this reason, some basic and fractal connection between human land-tenure systems and political economies and instinctive population dispersal may be inferred. These observations will underpin a theory to explore and explain questions about human population overshoot and different political systems and ways of distributing energy.

For the traditional sociologist who is trained not to compare humans to other species, this line of thinking may seem hopelessly radical. Remember, however, that it has long been a sociological tradition to jealously claim incest avoidance as something which distinguishes humans from 'lower' creatures. Do we now reject discussion of this phenomenon because it binds us so irrevocably to the rest of the animal kingdom? Or do we accept that incest avoidance as a dynamic in population spacing implies that there is potential for cooperation rather than for conflict?

Let us now proceed as if incest avoidance – both blood and 'in-law' - or population dispersal along principles evoked in the Westermarck effect - is the norm.

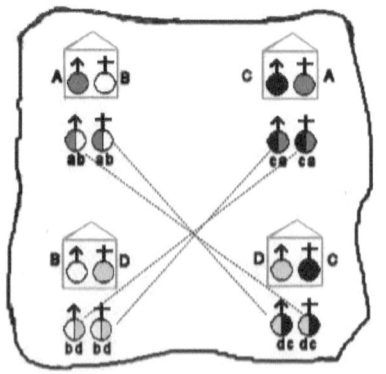 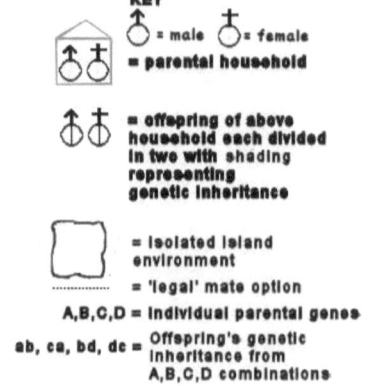

Fig 1. Organisation with incest avoidance of 50% and 25% on an imaginary isolated island among two generations with limited genetic diversity.

If we think about the patterns (beginning with the simple one in Figure 1) that result across a landscape and in time from incest avoidance, the results would be a binary system that grows in symmetrical geometric progression, according to landscape fertility and time. It would be organized fractally. This binary system defines the limits of fertility opportunity according to the norm of incest avoidance. In such a system, brother may not marry his sister, daughter or mother. Usually brother may also not marry grandmother, aunt or cousin or niece. Of course this also means that sister may not marry brother, son, father, grandfather, uncle, cousin or nephew. Siblings of the same two parents are related to each other at the rate of 50% each. Cousins, grandparents, nieces and nephews, aunts and uncles, are related by 25%.

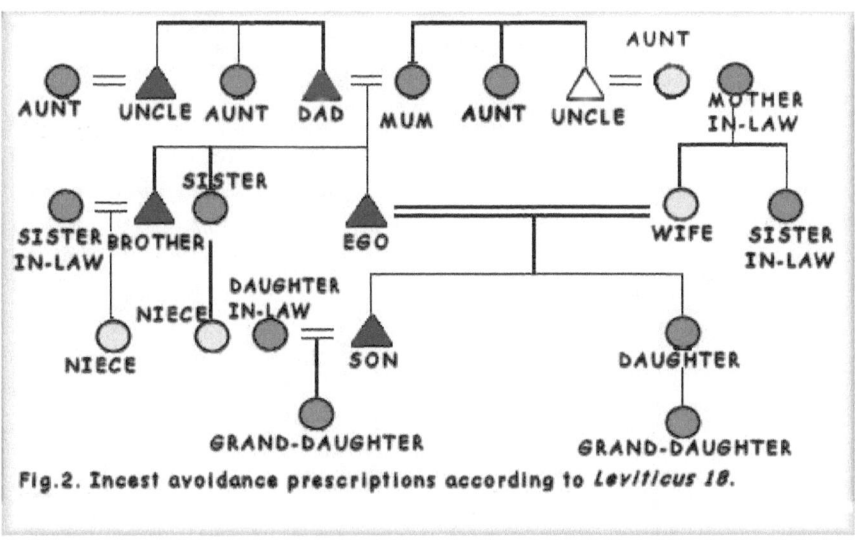

Fig.2. Incest avoidance prescriptions according to *Leviticus 18*.

Figure 2 is a geometric illustration of incest avoidance prescriptions according to *Leviticus 18*. 'Ego' is the central reference person. Patrilineally-related men are shown as triangles; excluded marriage partners are shown as dark circles; partners not explicitly excluded are shown as lighter circles. (The figure was adapted from diagrams by Brian Schwimmer.) [37]

Figure 2 above gives a good idea of the reduced range of opportunities to sire legal heirs in a typical, well-known and well-documented incest prohibition system, usefully documented in Leviticus 18 in the Bible.[38] Out of 16 women, only four were legally available.

Let us now consider, diagrammatically, what could happen to fracture this predictable arrangement. We now consider two secure populations with intact clans where incest avoidance and the Westermarck and in-law effect is maintained.

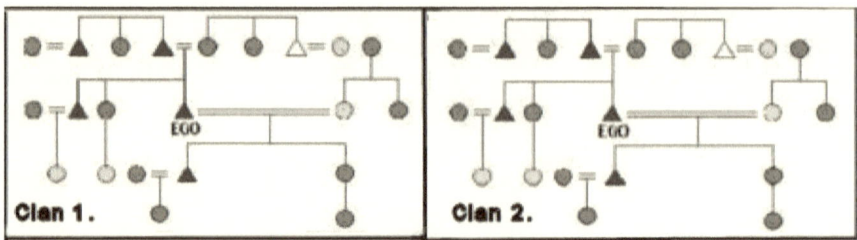

Fig. 3. Two clans, each on a separate island.

In Figure 3 each rectangle represents an island. Each island is inhabited by a small clan, called Clan 1 and Clan 2. 'Ego' in both clans may only father children by four out of 12 women in his clan. The only way that non-clan members can peacefully enter the other clan's area is via marriage or authorized immigration.

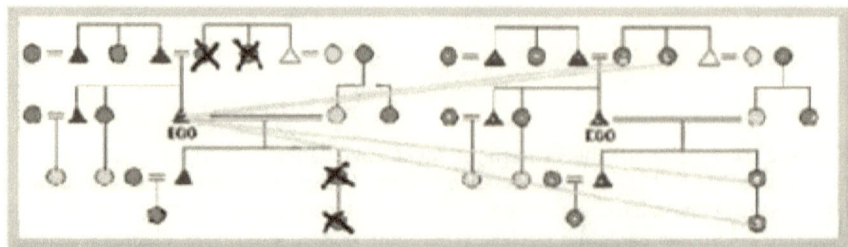

Fig.4. Violent merging of clans changes fertility options.

In Figure 4 above there has been a violent invasion. Four women have died in Clan 1. All of them were off-limits for Ego in Clan1. Now four women from the unrelated clan 2 in the next island have entered territory no.1.

Suddenly 'Ego' has four more potential partners with possibility of legally fathering children by 8 women. His chances of fertility are therefore doubled, although the population numbers and density of Territory no.1 have not changed. What has changed is the fertility of the population in territory no 1.[39] (Actually it is more likely that males would invade and the women would find partners.)

These patterns and their fracturing can be extrapolated to human clan and village organisation, and then on to bigger societies where, for a variety of reasons, clan-based sequences have been interrupted and new layers have been built on top of fragmented old layers, around 'energy gradients' like river deltas. (There are many historical and actual examples, such as the Fertile Crescent, the Ganges Delta, the Nile Delta, Rome, Cairo, London, Sydney etc.)

In societies that do not rely on artificially modulated environments, population density reflects the fertility of the local environment. My hypothesis is that the environment affects hormonal feedback and hormonal feedback affects the fertility of individuals. The fertility of individuals is physiologically, socially and organisationally modulated by incest avoidance and the Westermarck effect, which is carried out more or less strictly depending on this hormonal feedback. Such patterns of partnering and reproduction are articulated in human cultures as laws and traditions. Human societies and their politics also reflect these patterns in their use of submission and dominance (as loyalty, obligation and power) in social hierarchies.

In 'modern' societies, characterized by national, regional, local and international population movement, and rich and poor food-sources imported from distant places, these biological population-spacing mechanisms have little or no relevant local biophysical and interpersonal political interface. That is, people don't think about them much and it occurs to almost no-one (has it ever been written about before?) that the opportunities for productive marriage in vast modern populations where a substantial proportion are constantly relocating, are amazingly numerous and varied compared to those in earlier societies.

Early peoples were born into territorial patterns which reflected the binary oppositions implicit in blood and marriage relationships, intersected by landscape concerns. All simple and traditional societies perpetuate these forms in their land-tenure and land-use planning.

In its most basic form, incest avoidance decrees that children move some distance away from their parents before forming their own households, since each human requires a certain amount of 'social' territory to mature and differentiate from his/her parent, and 'economic' territory to survive. Usually one or the other sex will tend to go and live nearer their mate/spouse's family, whilst the other sex will bring their spouse/mate back.[40] In a patriarchal society, it is usually the woman who goes to live with the husband's family. And, when a clan or more complex society (such as a tribe or village) reaches a certain population density (anecdotally between 300 and 500 people in a clan) whereby the territory it requires to sustain its inhabitants starts to reach beyond comfortable daily travelling distances, a new clan or other settlement will form.

If there is no-one within the available area who is not closely related, then some individuals will remain without inheritors.[41] Or, if there is not enough territory, in a subsistence culture, pregnancy will be avoided, aborted or any children produced will be neglected, destroyed or adopted out where possible, sold into slavery or indentured as low-paid workers and servants.

The inference is that, before complex agriculturally based societies developed, human families formed clan and tribal populations, in densities which reflected local biophysical opportunity. Yet another way of saying this is that the density of these populations and the distance between settlements was a function of the binary opposition of blood and in-law incest avoidance and the Westermarck effect,[42] interacting with soil fertility, climate, and natural features. These factors are tempered by the kind of economy the society had. In a hunter-gathering or herding economy more 'horisontal' territory will be required per person than in an economy that relies successfully on intensive farming, which 'mines' vertical territory in soil. In the case of coal and oil-based economies, extractive methods access not only vertical territory, but the goodness of soils, plants and animals, stored from other times, and may support many more people per ha than other systems, subject to fuel exhaustion and pollution.

In traditional societies land was inherited. There was no system whereby parts of clan, tribe or village territory could be aggregated and sold off to local villagers or to outsiders. Since land could only be passed on through inheritance, the pattern of settlement preserved its binary arrangement around blood and marriage. This was a biological algorithm that functioned within the local and regional biophysical environment, according to physical laws like the rules of thermodynamics as they mediate life-support.

Such was the foundation of a steady state society, where population varied little and there was little difference between the land rights and status of individuals. Numerous variations were possible within this 'topology', for instance, it could be matriarchal or patriarchal, polygynous or polyandrous, but some things could break it. What can break this topology, changing systems within and outcomes, is the wider subject of my ongoing research and writing.

Culture and Structure

Humans are named according to their family lineage and their place of birth. Individuals may bear the name of their clan or tribe, which may also reflect the name of that tribe's territory, and special rights to hunt or obligations to avoid eating a particular animal. Such names are signposts to the blood or 'marital' relationships between the individuals of known clans and tribes in the area. Whereas names in a small stable society are like maps of local and regional genetic and ecological relationships, in many modern societies, names may have lost all intelligible relation to geography and clan. They are almost meaningless artefacts, except where they acquire new meaning through becoming attached to statistical data about individuals. Much of these data have arbitrary or specialised significance and are of little use to the individuals concerned.

Human cultures are quasi-transportable and this is one reason to explain why people transplanted to different new environments may behave in a counter-intuitive way. They are seeing their new environment through a cultural filter. The other qualities of dominance that express themselves in our family, clan, tribe and national politics tend to preserve established patterns. For instance, normal conformity to peer values and pressure makes it very difficult for most individuals to adapt independently to changes around them if the more dominant social forces seek to retain positions adapted to other times and climes. Peer pressure is a strong cement that complements the power of dominant males and females over their more submissive fellows. Such bonds are integral to the order and structure of societies. The bonds involve power attributes such as territorial rights and obligations which motivate the incumbents to maintain position. Cultures that use money have to deal with vast networks of invisible bonds which have largely lost their original ties to geographically based land and resources. The laws of cultures – particularly those of land-tenure, marriage and inheritance - reflect these kinds of arrangements and are used to bolster them. Culture is our chief means of future planning.

The complex internal rigidity of cultures is yet another reason why it is difficult for complex societies, where both environment and culture are artificially mediated, to adapt their land-use planning and their population regulators to new conditions where flatter smaller societies might be in order. A case in point would be the need to plan for global climate change and petroleum depletion which, it seems, is better met by older, less mutilated cultures like those of Western Continental Europe than by the Anglophone settler societies and other products of colonialism with their broken land-tenure topologies.[43]

Social Geometries

In a clan based society you would expect there to be a regular predictable geometric spatial relationship between persons, families and settlements, intersected by landscape and land fertility qualities. Habitations would remain with little change over generations, as they would be occupied by succeeding generations with few changes in numbers or needs.

Where clan series have been discombobulated, however, this geometric regularity would be broken. This breaking would then permit a new irregular mosaic pattern created through new incest avoidance and Westermarck effect patterns. Such disruptions are often detectable in villages and cities which have been subject to mass influxes, like settlement layers in archeological digs.

The mosaic would permit greater population density. Shanty towns and slums reflect this kind of demographic disorganisation in their high density and irregular agglutination. The practice of 'in-filling' - an Australian planning term to describe the re-zoning of built areas to higher densities - is another sign of demographic breaks in series.

Where original clan settlement patterns have been broken, memory has also been broken, so there would be incidental incestuous relationships due to new mosaic-pattern-dependent spatial proximity. Where memory has broken down, obligations, responsibilities and loyalties, are also forgotten and new ones must be forged. Communities are vulnerable during these periods.

Spatial relationships may also be modifiable by custom-defined gender pathways. An example might be the aboriginal custom of closing your eyes in order to avoid seeing a close relative of the opposite sex; gender practices involving the sexes spending much of their time working in different parts of the clan territory, or of one sex taking a very long route to do something which could be done via a much shorter route, but whereby the longer route keeps them away from the opposite sex; the practice of having separate houses and separate villages, separate land for the two sexes and those more familiar modern gender divisions, such as single sex schools and occupations that tend to be reserved for one sex or the other.

Some of us would also recognise that there has been a reduction in gender-specific customs in 'post-industrial' societies, with less gender specific entry into occupations and into institutions such as schools and churches. The 'uniforms' more familiar in 'mechanical' societies[44] where the range of head-dresses, scars and tattoos may be limited to one kind for each sex have been replaced by a host of different clothing styles, with elements of cross-dressing or of unisex apparel common.

The theory I have developed about incest avoidance and population dispersal does seem to have strong fractal elements. Australian environmentalist, Greg Wood, was reminded by my theorizing of the laws of thermodynamics and their impact on order. Remarking that, although most energy expenditure leads to disorder, life restores some order before it degenerates as well, he suggested that my theory of binary oppositions as a limiter of ultimate clan, tribe or city population size, might also apply to individual organs and to the phenomenon of cancer – whereby organ tissue multiplies and entrophies (becomes disorderly).

The notion of cellular division and then sexual differentiation are like fractals on a continuum, with cellular division in protozoa at one end and sexual dimorphism and incest and Westermarck avoidance, and separation between the discreet populations represented by clans, tribes, cities and nations at the other.

4. New Theory of Demographic Adjustment

I am aware of no other sociological theories which compare the effects on population size of different systems of land-use planning, marriage and inheritance. This is despite the fact that my theories derive from observations which were the building blocks of much anthropology prior to the 1980s. The material I use to identify algorithms of population spacing in other species is recent and was developed for other purposes, mostly to establish whether incest avoidance occurs in non-human species (a very rich area), and then to measure its effect on genetic health.[45]

Perhaps due to the assumption that genetically programmed incest avoidance has evolved as a norm because it diminishes the risk of deformities and other inherited problems, the effect that avoiding incest has on territorial division and population spacing – i.e. on population growth and its spatial distribution – seems so far to have gone unremarked in these research conclusions. We read that incest avoidance is achieved by dispersal but this is not turned round to observe that incest avoidance **causes** dispersal.

In my theory, if algorithms of population spacing occur in other species they must be biologically and genetically based. I then hypothesize that incest avoidance norms in our own species almost certainly arise from similar autonomic sources. As I have suggested, in studies of other species, there is good evidence that these algorithms adjust to hormonal responses to sensory and alimentary feedback from the environment where availability of territory may be measured by distance from close relatives. Potentially fertile offspring who are dominated by close relatives cannot find territory of their own and non-incestuous, non-Westermarck effect opportunities for mating are not taken up. There is evidence that hormonal suppression also occurs in similar circumstances in humans. For instance it has been observed that girls raised in the presence of stepfathers menstruate earlier than those raised in the presence of their blood fathers.[46] Incest is more likely to occur in such cases as well, in ours and other species.[47]

In human societies, indications of incest avoidance as a major factor in population spacing are most obvious in rules for marriage and inheritance laws where marriage is defined as a union which gives rise to legitimate children who have family rights to inheritance and, in modern societies, citizens' rights to land or an equivalent, such as employment or unemployment benefits, or public housing, and the vote and a passport. In a subsistence society a legitimate child is born with land rights; illegitimacy equates to being born in a situation that does not entitle the person to land. Not having land-rights means that the person has no means to survive independently. Societies have evolved where the land-tenure system allows legitimate children to be disinherited. (The great Anglophone societies, starting with the UK, are products of this system, which they have transmitted to many colonies. It is the basis of private property and of capitalism.[48]) In such cases the State may accord them certain rights of citizenship, often of a marginal variety. Industrial societies **rely** on a working class which lacks sufficient land-rights for independent self-support.

These differences in social structure and function are well illustrated below by the authors of an introductory unit for the course, "Aboriginal cultures and the land".

> "[...] because Aboriginal societies were highly egalitarian, there was/is never any need to produce 'surpluses' so that the labour of many could/can sustain the wealth of a few—a primary characteristic of what we usually call 'civilisation', where oppression of both our fellow humans and of the natural world are fundamental to what passes for 'civilised' society. [Add slavery, widespread conquest by warfare, writing, and building in stone, and you get a 'Great Civilisation', such as ancient Rome or China!]"[49]

The same introductory text affords us a good compare and contrast between a society structured primarily by kinship and one commercially structured. The theoretically almost unlimited choice of partners in 'Western' societies, compared to the very limited choice in aboriginal cultures, is also remarked upon:

> "1–15
>
> Kinship in Western society tends to be fairly narrowly defined. The isolated nuclear family is of great significance; other relatives

may also be important in a secondary way, but in general this also involves a restricted range of people.

Many aspects of social life are organised through other institutions—the economy, the polity, the education system, etc.; kinship is limited to a particular sphere of social life (the family), rather than being a pervasive structuring principle as it is in many societies.

Western societies emphasise patrilineal descent. Despite some recent changes, names are allocated patrilineally. In the past there was also a male bias in the inheritance of property. [Note that this was not the invariable case throughout European cultures – SMN.]In some cases there may still be traces of this when sons rather than daughters inherit the family trade, farm or business.

'Coming from a good family' can still be important, but this is not highly structured in terms of particular, defined lineages (it was much more important in the past, when the monarchy and the aristocracy were of greater significance).

Cross cousins and parallel cousins are not distinguished. Residence after marriage is neolocal. Exogamy only applies to a small range of close kin (those to whom the incest tabu applies). Apart from this the choice of marriage partners is in theory unlimited—there is no formal system of endogamy. Informally, however, people do tend to marry those of their own ethnic and socioeconomic status.

Generally, kinship and descent tend to be overshadowed by other factors—formal education, the bureaucratic organisation of work, job mobility, social security, institutionalised health care systems, and so on. Apart from the nuclear family, groups based on kinship or descent do not play a large part in the organisation of [Western]society."[50]

All human land-use planning, marriage and inheritance laws explicitly incorporate incest avoidance to some degree or other, including explicit exceptions.[51] The permissible variations on incest avoidance have been iconicised in religion and codified in law in every society.

Institutionalised Exceptions to Incest avoidance affecting land-tenure

There are exceptions to nearly every law and incest avoidance has its exceptions as well. Where these are institutionalised they always seem to be related to securing land and other power bases.

The Bible has already been mentioned as a source of commonly prescribed marriage rules. A widespread example in some societies is the 'levirate', in which tradition expects a man to marry his deceased brother's widow if they live on the same estate and the deceased has no male heir. This keeps the land in the same family lineage and any new descendents may be treated as children of the deceased.[52]

Another form also prescribed in the Bible was to marry daughters to their father's brother's sons (the girl's first cousins). This was also a way of preserving land in the same male line because the male cousins descend from the same grandfather. This is known as 'parallel cousin marriage and lineage endogamy'.[53]

Even slaves (presumably as family possessions) would sometimes be coopted as legal spouses for the purpose of securing an heir.

Anthropologist Brian Schwimmer,[54] provides a useful visualisation of these arrangements, which I have copied below with modifications for black and white print..

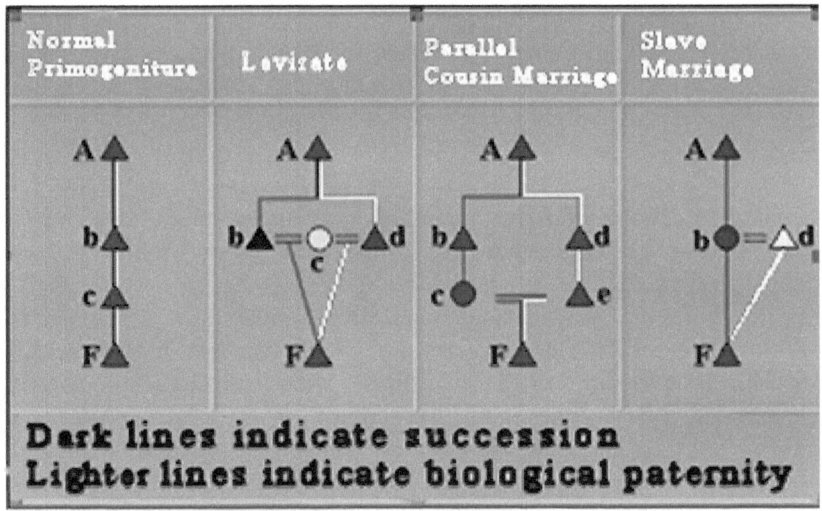

Normal Primogeniture	Levirate	Parallel Cousin Marriage	Slave Marriage

Dark lines indicate succession
Lighter lines indicate biological paternity

The case of Egyptian pharaohs where brothers and sisters were married, is well known. First degree incest may also occur through the union of closely related cousins in caste intermarriage, as practised among royalty. Caste systems also evolve in colonies, where the colonisers preserve land-acquisitions, power and solidarity, by excluding the colonised as marriage partners. Ancient Greek colonists in Africa married brother to sister and, in the Norman class in England from 1066, caste separateness was the structural underpinning of the feudal system of land-aggregation and the social structure of work.

Domestic animals in modern human societies

Exceptions to the incest and Westermarck effect are also observable in the behaviour of domestic animals that live in human societies. These animals tend to practice instinctive incest avoidance when raised normally in the wild. For instance, the normal familial society of wolves contains one breeding couple closely associated with non-breeding related males and females in a close-knit extended family.[55] But, among the descendents of wolves – domestic dogs – the clan arrangements are rarely preserved and dogs often don't know who their fathers, sisters, cousins and uncles are, due to the dispersal of litters. Such animals have more opportunities to mate because the order of the clan is absent. There is similar social fragmentation and loss of kinship knowledge in countries which are 'melting pots' or war-torn, where people are subject to frequent geographical and social dislocation. The children of artificial insemination, be they human or cattle, have no chance of consciously avoiding blood-relation incest since they don't know their fathers, siblings or uncles and aunts.

Runaway populations of felines, canines and humans may have similar causes. In the 'settler societies' of the United States, Australia, New Zealand, and Canada, indigenous populations have been greatly disorganised. So, however have the settler populations, perhaps even more so. The result is a chaotic soup where many do not have any relationship with the land or the incumbents with which to orientate their incest avoidance or Westermarck gyroscope. Furthermore, territory is defined by contract and acquired through money, not through birth-right. What effect must this have on the innate skill we infer here to be normally available to members of populations in stable situations to gauge safe levels of reproduction, well within carrying capacity and comfortable density?

Some other questions arising

Kinship theory has demonstrated that humans and other animals are more careful of the feelings and rights of their blood relatives and clans. I think that we have to consider the application of the Westermarck Effect here and consider that it might be better to restate this as that humans and animals are more careful of the feelings and rights of those they grew up with. **If the ties to these people don't exist, or new Westermarck ties are not mediated by a traditional biological framework of dominance,** what effect might it have on justice and equity overall? Even if we have a sociable concern for our neighbours, what is the effect of competition when the loss of our kinship-territory gyroscopes causes situations where we are forced to compete for scarce land, resources such as water, and services – such as higher education, hospital beds, and well-paid employment? It was this kind of disruption of kinship organisation and loyalties that permitted the formation of feudal system and industrial society with its stark social division of work and property ownership.

These risky situations are not confined to the 'developed' Anglophone societies; they appear to be at the roots of the chaos in many other societies, for instance, in those Pacific, African, Indian countries, states and regions, where dispossession has been profound, sustained and repeated. The situations in 'developing' countries which have managed to preserve kinship series and their relationship to land is usually better. China is somewhat better off than India, albeit China suffered profound dislocation by Mao and subsequent post revolution regimes; Japan, which might have been overwhelmed by war and colonisation, has retained many kinship series and the related aspects of land-tenure sufficient to preserve its forests and to be among very few nations with islands in the Pacific that do not suffer from overpopulation; its citizens have achieved higher standards of living than most Europeans.

The relationship with the land is our normal means of territorial feedback regarding available necessary space from which to derive a reasonable living. In the Anglophone first world countries where populations are growing rapidly, this normal relationship has been replaced by abstract environmental feedback, via media, opinion, arbitrary statistics and political marketing. As suggested by research cited earlier in this book,[56] underpinning all environmental feedback is probably reliability of nourishment. International trade has obscured seasonal variations in available foods and abundant fossil fuel has removed seasonal and regional limitations to available calories. One result of this is a much greater intake of calories, particularly of rich fats, in some first world populations. An impact of this increased fat intake has been obesity and decreased age of menarche. Another impact, also apparently acting on hormonal feedback, has been a rise in diabetes. This implies a fall in life expectancy. Who knows what the ultimate equation might be in response to the chemical soup of the massive populations of the 21st century?

5. How disruption of the binary incest avoidance system would deliver more fertility opportunities

How would the disruption of the binary incest avoidance system work to deliver more fertility opportunities? Consider that, where populations of any animal are destroyed and families and clans dispersed, parents become separated from children and sibling from sibling. Sometimes entire peoples perish. If there are survivors though, under such confused conditions, they grow up without knowing their parents, their siblings or their ancestral lands. Land-rights are left undefended and now people from other places may move in and use the land differently.

Disruption could be so drastic that, if the opportunity arises to procreate with an unrecognised parent or sibling, or 'in-law', no incest avoidance or Westermarck response, which in humans requires close association during the first 30 months of life, would have been initiated.[57] The spatially organizing parental and other blood relative dominance would be missing.

Fertility opportunities may thus be dramatically increased after an initial die-off since individuals may be thrown together in groupings and in new places where every individual is a stranger and therefore all those of the opposite sex could be legitimate potential mates. The stress for the displaced and harassed to organize through bonds with others of their kind would be a natural survival strategy. The result could be a population explosion – in humans, kangaroos or any species which can normally be shown to practice incest avoidance. This could explain the extremely high rates of fertility in troubled African states, for instance.

This concern may seem exaggerated to the modern reader if they come from one of the many structurally disorganized modern societies, where concern for kinship has a low priority in the economic order that rules those societies. This is particularly true of the Anglophone settler societies. In such countries clan organization has been obscured, fragmented and subsumed to major and minor transmigrations. People locate away from their home-towns to be close to employment and form routines and attachments that bind them to new places and organizational structures (such as industrial workplaces) and alienate them from their roots. They marry people who come from completely different backgrounds, with different values and loyalties. Christmases are famously times of existential angst where sons and daughters go 'home' to places and people whose lives are completely separate from theirs. International and inter-regional migration is facilitated by cars and planes and dictated by economies where employers reserve the right to demand that spouse relocate with corporation. Spouses divorce if their loyalties to or dependencies on different employers over-ride their commitment to marriage. Ownership of land and ties to place are a hindrance to mobility in such cases as are the related ties of marriage. Renting, divorce and serial families become normative.

In a society where there is constant relocation within a huge population in a large territory, the probability of accidental incest seems so low that the risk arouses little or no concern. Rituals of avoidance beyond the primary family in a confined household are quasi-nonexistant. If there were concern it would be for the possibility of producing genetically compromised children. Modern society provides a technological solution for this though. Where individuals from an identifiable or self-identifying ethnic group or who believe they are cousins contemplate marriage, blood tests may be carried out to see if either is a carrier of an inheritable disease with high prevalence in that group – such as sickle cell anemia in descendents of Mediterranean peoples. The potential parents may then be counseled not to have children, or to have amniocentesis and consider aborting any affected fetuses.

As I have stated, the concern I am flagging here though is not of inherited diseases, but about unintentionally or unwisely multiplying the rate of population growth beyond safe levels, in terms of equity and environment. In fact unstoppable population growth and overshoot of resources, such as water, along with homelessness, are already fact in those huge modern first-world Anglophone societies that are constantly in structural flux.

This seems to be what happens in all disrupted populations - from kangaroos to cats, from mice to men.

About this book and future projects by the author

This book was about a new theory of population which is based on genetically based algorithms and food supply. It is the basis for another book in progress where the new theory is employed to re-examine history with particular reference to theories of societal collapse and the development of modern expansive and conservative economies and populations. The effect of two different land-tenure systems on the origins of industrial capitalism in Britain and modern democracy in France will be explored.

Sheila Newman

Correspondence to: astridnova@gmail.com

Blog: http://candobetter.net/SheilaNewman

REFERENCES

Abbott, D. H., Saltzman, W., Schultz-Darken, N. J., & Tannenbaum, P. L. (1998), "Adaptations to Subordinate Status in Female Marmoset Monkeys," *Comparative Biochemistry and Physiology*, 119, 261-274.

Abbott, D. H.,Saltzman, W., Schultz-Darken, N. J., & Smith, T. E. (1997), "Specific Neuroendocrine Mechanisms Not Involving Generalized Stress Mediate Social Regulation of Female Reproduction in Cooperatively Breeding Marmoset Monkeys," in *The Integrative Neurobiology of Affiliation*, 807, 219-238.

Abernethy, V., *Population Politics*, Transaction Publishers, Plenum Press, New York, (2005)

Andrewartha, H.G., and Birch, L.C. , *The Distribution and Abundance of Animals*, University of Chicago Press, Chicago, (1954), cited by David Pimentel in "Population Regulation and Genetic Feedback," *Science*, Vol.159, 29 March 1968, p.1433.

Author unknown, "Aboriginal cultures and the land: An introduction to the unit," Course at University of S.E. Queensland, www.humanities.cqu.edu.au/abtorres/52246/52246sg., p.73. (Accessed 2005)

Baker, J. V., Abbott, D. H., & Saltzman, W. (1999), "Social Determinants of Reproductive Failure in Male Common Marmosets Housed with their Natal Family," *Animal Behaviour*, 58,501-513.

Birdsell, Joseph, B., "Australia: Ecology, spacing mechanisms and adaptive behaviour in aboriginal land tenure", in Ron Crocombe, (Ed.), *Land Tenure in the Pacific*, OUP/MUP (1971), pp.334-361

Caughley, G. and Sinclair, A. R. E., Wildlife Ecology and Management, Blackwell Science: Boston, (1994), cited in Fletcher, D., "Population Dynamics of Eastern Grey Kangaroos in Temperate Grasslands" (PHD thesis), Inst of Applied Ecology, University of Canberra, http://erl.canberra.edu.au/uploads/approved/adt-AUC20070808.152438/public/02whole.pdf (2006)

Cockburn, A, Osmond, H, Mulder, R, Green, J, Double, C.(2003) *Divorce, dispersal and incest avoidance in the cooperatively breeding superb fairy-wren Malurus cyaneus,* in Journal of Animal Ecology, Volume: 72, Pages: 189-202

Charpentier, M., Peignot, P., Martine Hossaert-McKey, M., Gimenez,,O., Setchell, J.M. and Wickings, E.J., "Constraints on control: factors influencing reproductive success in male mandrills (Mandrillus sphinx)", CEFE-CNRS UMR 5175, 1919 Route de Mende, 34293 Montpellier Cedex 5, France, UGENET, CIRMF, Franceville, Gabon, and Department of Biological Anthropology, University of Cambridge, Cambridge, UK, [Behav Ecol 16:614–623 (2005)] *Behavioral Ecology,* doi:10.1093/beheco/ari034, Advance Access publication 23 February (2005)

Chepko-Sade, B.D., and Tang Halpin, Z., *Mamalian Dispersal Patterns,* Chicago Press, Chicago, 1987

Daadoun, Roger "Les régulations sociales de la sexualité", *Encyclopedie Universalis,* Electronic Edition. (1999)

Dallwig, Rebecca, [Summary entitled] "The Common Marmoset, Callithrix jacchus, Current Research," Page 8, III. Biomedical Research, in Primate Info Net, Library and Information Service, National Primate Research Center, University of Wisconsin – Madison, http://pin.primate.wisc.edu/callicam/research8.html (last accessed 1 Nov 2009)

Darwin, C., *The Origin of Species,* (1859) cited by Pimentel, D., in "Population Regulation and Genetic Feedback," *Science,* Vol.159, 29 March 1968, p.1433.

Diamond, J., *Collapse,* Penguin Books, (2005) p116.

Durkheim, E., *La division sociale du travail,* Paris, (1893)

Ekblom, Robert, (2000) "Inbreeding avoidance through mate choice",Introductory essay, Evolutionary Biology Centre, Department of Population Biology, Uppsala University, Norbyvägen 18 D, SE-752 36 Uppsala, Sweden, http://www.ebc.uu.se/popbio/people/rekblom/Ekblom%202000.pdf

Engelstädter, Jan and Charlat, Sylvain, "Outbreeding selects for spiteful cytoplasmic elements," *Proceedings of the Royal Society, Biological Sciences*, 2006 April 22; 273(1589): 923–929, Published online 2006 January 17. doi: 10.1098/rspb.2005.3411.

Ferguson, F., *Look Down, see the women cry*, from a Karen folktale he recorded as "7. The story of Nauj Htof K'Maiz and Cau Seif Laf Geiz, (Taj Leplez Nauj Htof K'Maiz dauv Cau Seif Laf Geiz,)" available in *Folktalkes/tajleplez*.pdf, archived at http://tonbo80.spaces.live.com/

"Frequently asked questions," Easter Island Foundation, www.islandheritage.org/faq.html (accessed 2005)

Fletcher, D., "Population Dynamics of Eastern Grey Kangaroos in Temperate Grasslands" (PHD thesis), Inst of Applied Ecology, University of Canberra. http://erl.canberra.edu.au/uploads/approved/adt-AUC20070808.152438/public/02whole.pdf (2006)

Griffin, AS., Pemberton, JM., Brotherton, PNM., McIlrath G., Gaynor D., Kansky R., O'Riain J., Clutton-Brock TH., (2003). A genetic analysis of breeding success in the cooperative meerkat (*Suricata suricatta*), Behav Ecol 14:472–480.

Hoier, S., "Father absence and the age of menarch, A test of four evolutionary models," (2003), *Human Nature*, Vol. 14, No. 3, pp. 209–233, Walter de Gruyter, Inc., New York.

Hopfenberg and Pimentel, "Human Population Numbers as a Function of Food Supply", 1 Duke University, Durham, NC, USA; 2 Cornell University, Ithaca, NY, USA http://panearth.org/WVPI/Papers/HumanPopulationNumbers.pdf (March 2001)

Hone J., Sibly, RM., "Demographic, mechanistic and density-dependent determinants of population growth rate: a case study in an avian predator," in 1: *Philos Trans R Soc Lond B Biol Sci.* (2002) Sep 29;357(1425):1171-7. Applied Ecology Research Group, University of Canberra, Canberra 2602.

Hugh, E., "Rethinking the demographic transition", www.edwardhugh.net/rethinkingsummary.html (No longer accessible.)

Jarne P, Charlesworth, D., "The Evolution of the selfing rate in functionally hermaphroditic plants and animals," *Annu. Rev. Ecol. Syst,.* 1993; 24:441–466. 10.1146/annurev.es.24.110193.002301

Koenig, W., and Haydock, J., in a website report on the "Social Behavior of the Cooperatively Breeding Acorn Woodpecker," Hastings Reservation and Museum of Vertebrate Zoology, University of California Berkeley, http://www.hastingsreserve.org/AcrnPkrs/AcrnPkrs.html (Accessed 2004) Now at http://www.hastingsreserve.org/Resident%20Web%20Pages/Koenig%20Web%20Pages/AWIntroPoster/AWposter.html. (Study began in 1971) (No date). Accessed 2 November 2009

Krebs, C. J. Two paradigms of population regulation, *Wildlife Research* **22**, 1-10, (1995) cited in Fletcher, D., "Population Dynamics of Eastern Grey Kangaroos in Temperate Grasslands" (PHD thesis), Inst of Applied Ecology, University of Canberra. http://erl.canberra.edu.au/uploads/approved/adt-AUC20070808.152438/public/02whole.pdf (2006)

Levi-Strausse, Claude, *Elementary Structures of Kinship,* (*Les structures élémentaires de la parenté*) Mouton et Maison des sciences de l'Homme, Paris, La Haye, 1967 (Ist édition, 1947)

Lihoreau, M. Zimmer, C. Rivault, C. , "Kin recognition and incest avoidance in a group-living insect," *Behavioural Ecology*, Vol.18, No.5, 2007, pp.880-887

Lukas, D., Reynolds, V., Boesch, C. and Vigilant, L., "To what extent does living in a group mean living with kin?", Max Planck Institute for Evolutionary Anthropology, Deutscher Platz 6,Leipzig 04103,Germany,†Oxford University,School of Anthropology,51,Banbury Road,Oxford OX2 6PE,United Kingdom, *Molecular Ecology* (2005)14 ,2181 –2196 doi:10.1111/j.1365-294X.2005.02560.x

Lutz, W., Testa, M.R., and Penn, Dustin J., "Population Density is a Key Factor in Declining Human Fertility," *Population & Environment*, March (2007)

Mateo, JM., (2003). Kin recognition in ground squirrels and other rodents. *J Mammal* 84:1163–1181.

Mowat, F., *Never Cry Wolf*, McClelland and Stewart, 1963

Newman, S.M., "Thomas Malthus and Australian thought," *The Social Contract*, Volume 8, Number 3 (Spring 1998)
http://www.thesocialcontract.com/artman2/publish/tsc0803/article_745
.shtml.

Newman, S.M., *The Growth Lobby and its Absence, in Australia and France*, Research Thesis, Swinburne University, (2002)
http://adt.lib.swin.edu.au/public/adt-
VSWT20060710.144805/index.html

Newman, S.M., "Land and Housing prices and Land-use planning and Housing Systems in Australia and elsewhere and the Impact of Globalisation, the Internet, Trends in Natural Increase and Immigration," Victorian Sustainable Population Australia submission to an Australian Housing Affordability Inquiry, (2003),
http://www.pc.gov.au/__data/assets/pdf_file/0020/58115/sub153.pdf.
The inquiry itself was conducted in 2003 and was called, "First Home Ownership Public inquiry."

Newman, S.M., "ACT Roo killings: Who profits? Behind the Earless Dragon mask," May 25, (2009). http://candobetter.org/node/1274

Pimentel, D., "Population Regulation and Genetic Feedback," *Science*, Vol.159, 29 March (1968), p.1433.

Pusey, A., Wolf, M., (1996). Inbreeding avoidance in animals. *Trends Ecol Evol* 11:201–206;

Shepher, J. *Incest, A biosocial view*, Academic Press, New York, (1983)

Stow, A.J., Sunnucks, P., (2004). "Inbreeding avoidance in Cunningham's skinks (*Egernia cunninghami*) in natural and fragmented habitat," *Mol Ecol* 13:443–447.

Yu XD, Sun RY, Fang JM, 2004. Effect of kinship on social behaviors in Brandt's voles (*Microtus brandti*)," *J Ethol* 22:17–22.

Schwimmer, Brian, "Hebrew social organisation," Department of Anthropology, University of Manitoba, (1995).
http://umanitoba.ca/faculties/arts/anthropology/tutor/case_studies/he

brews/marriage.html#levirate. (Accessed 2 November 2009). Permission to use diagrams has been obtained.

Waldman, Bruce, Rice, John E. , Honeycutt, Rodney L., "Kin Recognition and Incest Avoidance in Toads," *Integrative and comparative biology*, Volume 32, Number 1, pp. 18-30.

Wilson, E.O., "Nature Matters", *American Journal of Preventative Medicine*, (2001), Apr;20(3):241-2.

ENDNOTES

[1] **Explanation of some terms:**

"Algorithm": A biological algorithm would be an inherited response tailored to an internal or external situation. The term is used by E.O. Wilson, "Instinctive habitat selection is universal, and its analysis has become an important industry within the growing discipline of behavioral ecology. Even tiny species of insects, rotifers, and other invertebrates, whose brains are invisible to the naked eye, follow impressive algorithms of orientation to reach the habitats they need to survive." He also uses the term, 'epigenetic rule', "defined as an inherited regularity of development." He gives incest avoidance and the Westermarck effects as examples of epigenetic rules. "Westermarck effect is an example of an epigenetic rule, defined as an inherited regularity of development." Source: E.O. Wilson, "Nature Matters", *American Journal of Preventative Medicine*, (2001), Apr;20(3):241-2.:

Some other examples of epigenetic processes are heritable changes in gene function that occur without a change in sequence of DNA. Two examples are 'X inactivation' where one X chromosome becomes inactive in females, and 'gene silencing'.

[2] Daadoun, Roger, (1999).

[3] (Discussed further on in this book.) This research is summarised in: Dallwig, Rebecca, The Common Marmoset, Callithrix jacchus, Current Research, Page 8, III. Biomedical Research [cont.], http://pin.primate.wisc.edu/callicam/research8.html (accessed 1 Nov 2009) and appears in many articles, including:

Abbott,D. H., Saltzman, W., Schultz-Darken, N. J., & Tannenbaum, P. L. (1998);

Abbott, D. H.,Saltzman, W., Schultz-Darken, N. J., & Smith, T. E. (1997);

Baker, J. V., Abbott, D. H., & Saltzman, W. (1999).

[4] Except, for instance, where genes causing problems do not manifest until after reproductive maturity. An example of this would be Huntingtons Chorea.

[5] Levi-Strausse, C., (1949)

[6] *Ibid*

[7] Shepher, J. (1983)

[8] The application of this theory to the Easter Island collapse and the history of capitalism in Britain and to Democracy in France is the subject of a book in progress.

[9] Waldman, Bruce, Rice, John E., Honeycutt, Rodney L., (2007)

[10] Jarne P, Charlesworth D., (1993)

[11] Engelstädter, Jan and Charlat, Sylvain, (2006)

[12] See, for instance, Fletcher, D., (2006) "A third approach explicitly involves more than one trophic level, by relating population growth rate to an ecological factor such as food availability, thus conforming to the 'mechanistic paradigm' (Caughley and Sinclair 1994; Krebs 1995, 2002). This is the least commonly adopted conceptual approach but it is the one underlying this study."

[13] Pimentel, D, (March 1968)

[14] Andrewartha, H.G. and Birch, L.C., (1954), and Darwin, Charles, (1859) in Pimentel, D. (March 1968), p.1433.

[15] Pimentel, D. (1968) p. 1434.

[16] Hopfenberg and Pimentel, (March 2001)

[17] Op.cit., p.4.

[18] Op.Cit. p.4.

[19] Sheila Newman, "Thomas Malthus and Australian thought," (Spring 1998). Early explorers and economists also knew that the more a place looked like a desert, the less population it would support. Joseph Banks, for instance, on the first voyage of Captain Cook to Australia, remarked on the low density of the indigenous population and assumed that this was a reflection of the low fertility

of the land. Coming from overcrowded Britain, however, he could not understand how the Aborigines kept their population density down. Thomas Malthus referred to Joseph Banks's comments (confusing them with Captain Cook's) in his second book on population.

[20] Lutz, W., Testa, M.R., and Penn, Dustin J., March (2007):

"Reproduction has been found to decline with increasing population density in a wide variety of species, yet demographers have not given systematic attention to density as a relevant factor in human reproduction".

Although density variations carry their own explanation in low environmental fertility, this self-evidence seems to escape many modern planners and economists who, seeing a sparsely settled area will advocate that it be put to more intensive use.

[21] Hone J, Sibly RM. (2002) "The negative relationship between population density and population growth rate is at the heart of population biology."

"Identifying the determinants of population growth rate is a central topic in population ecology. Three approaches (demographic, mechanistic and density-dependent) used historically to describe the determinants of population growth rate are here compared and combined for an avian predator, the barn owl (Tyto alba). The owl population remained approximately stable (r approximately 0) throughout the period from 1979 to 1991. There was no evidence of density dependence as assessed by goodness of fit to logistic population growth. The finite (lambda) and instantaneous (r) population growth rates were significantly positively related to food (field vole) availability. The demographic rates, annual adult mortality, juvenile mortality and annual fecundity were reported to be correlated with vole abundance. The best fit (R(2) = 0.82) numerical response of the owl population described a positive effect of food (field voles) and a negative additive effect of owl abundance on r. The numerical response of the barn owl population to food availability was estimated from both census and demographic data, with very similar results. Our analysis shows how the demographic and mechanistic determinants of population growth rate are linked; food availability determines demographic rates, and demographic rates determine population growth rate. The effects of food availability on population growth rate are modified by predator abundance.

[22] Hopfenberg and Pimentel, "Human Population Numbers as a Function of Food Supply", (2001), p.4:

"The finding that the population size of animal species is a function of food availability has been empirically demonstrated. Food energy is partitioned into four compartments viz.: maintenance, growth, stored energy, and reproduction. Scott and Fore (1995) investigated the effects of food availability on reproduction in the marbled salamander. Subjects were assigned to one of three groups. At the end of the experiment, 60% of the high-food females were reproductive. In the medium-and low-food groups, these numbers were 42% and 12% respectively. These results demonstrate that food availability influences the population dynamics of a species.

Similarly, Komdeur (1996) demonstrated that the Seychelles warbler prolonged their reproductive season, including increases to year-round breeding, when their natural condition changed to one with high food availability. Conversely, in female musk shrews (whose sexual receptivity is not restricted to the preovulatory period), 48 h of food restriction led to reduced mating behavior compared with ad-lib controls.

Thus, small reductions in food availability can inhibit female sexual behavior (Gill and Rissman, 1997). In the Calanus finmarchicus, egg production is suppressed when the nutrient pool decreases below a minimal critical value. There-after, no eggs are laid. When food is reintroduced, somatic growth resumes until structural body weight is restored, then oogenesis is fueled (Carlotti and Hirche, 1997). Also, Iwamoto (1978) has shown that monkey troop size increases rapidly after artificial provisioning, but the level of consumption efficiency of the troop is always maintained lower than the critical point in both the artificial and natural habitat. Starvation within the troop simply does not occur if the rate of food avail-ability is held relatively constant. Under natural conditions, as the feeder population increases, the food population decreases. This leads to a decrease in the feeder pop-ulation which is then followed by an increase in the food population. This increase in food availability again produces an increase in the feeder population. In quaternary consumer species, the so-called 'top of the food chain', this occurs primarily through fluctuations in birth rates.

[23] Hopfenberg and Pimentel, "Human Population Numbers as a Function of Food Supply", (2001) p. 4.

[24] Joseph B. Birdsell, (1971), pp.334-361. Birdsell suggested it: "Tindale's (1940) map of the territorial boundaries of approximately 400 dialectical tribes reveals an orderly spacing of people throughout the continent, and although the intensity of land use varied dramatically with changing environments, there were no empty or unclaimed spaces.

From Tindale's basic data Birdsell (1953) produced a quantitative ecological analysis which demonstrated the importance of a single environmental variable, mean annual rainfall, in determining the size and pop density of tribal areas. A single enviro determinant cannot completely express the relationship of the numbers of men to their land, but in this case a coefficient of cuvilinear correlation of 0.18 was obtained. This close relationship was determined for basic ecological units, the 123 dialectical tribes whose resources primarily depended upon locally earned rainfall. The mathematical function of the relationship was logarithmic and areas decreased and densities increased as the mean annual rainfall rose. The analysis showed continuous variation, as opposed to stepped variation, between the minimum rainfall of four inches, through the total range to the maximum of 160 inches annually in the rain forests of north-east Queensland."

[25] Robert Ekblom, "Inbreeding avoidance through mate choice", Evolutionary Biology Centre, Department of Population Biology, Uppsala University, Norbyvägen 18 D, SE-752 36 Uppsala, Sweden, robert.ekblom@ebc.uu.se:

"4.2 Reproductive suppression and delayed maturity One way of dealing with inbreeding depression in animals that live in family groups, is delayed sexual maturation (or other forms of suppression of sexual activity) when the parent of the opposite sex is still present in the group (Pusey and Wolf, 1996). This phenomenon is known in prairie dogs (Cynomus ludovicianus), where a young female is significantly less likely to come into oestrus when her father is still living in her natal coterie territory (Hoogland, 1982, 1992). The mechanisms behind this are, as yet, unknown but one possibility is that hormonal activity could be affected by the scent of related individuals (Blouin and Blouin, 1988). In the communally nesting acorn woodpeckers (Melanerpes formicivourus) a female does not breed in her natal territory until her father has been replaced. Other females of the same age, however, reproduce after having migrated out from their natal group (Koenig and Pitelka, 1979). No true assessment of genetic similarity is needed for this kind of incest avoidance. Individuals must, however, be able to identify different individuals of their family and some kind of regulation of reproduction in response to this is necessary."

[26] Ekblom, R. (No date): "It is important to note that mate choice and dispersal are non-exclusive to one another. Instead, one common driving force of male dispersal is competition for mates. If females prefer to mate with unrelated males, then males should profit from dispersing out of the natal territory to find unrelated females. Such female mediated male migration has been found in for example olive baboons (Papio anubis) (Packer, 1979) and white-footed mice (Peromyscus leucopus) (Keane, 1990). In this way, female mammals could force males to bear the high cost of inbreeding avoidance (dispersal) by using a cheap mechanism (choice). In birds, females are usually the dispersing sex, a fact that

has puzzled investigators of inbreeding avoidance (Pusey, 1987). If the costs of dispersal are smaller in birds than in mammals, this could possibly lead females to disperse themselves instead of forcing the males to disperse by means of mate choice."

[27] E.O. Wilson, (2001)

[28] Koenig, W. and Haydock, J., "Social Behavior of the Cooperatively Breeding Acorn Woodpecker." (No date)

[29] Cockburn, A, Osmond, H, Mulder, R, Green, J, Double, C.(2003)

[30] Marie Charpentier, Patricia Peignot, Martine Hossaert-McKey, Olivier Gimenez, Joanna M. Setchell, and E. Jean Wickings., "Constraints on control: factors influencing reproductive success in male mandrills (Mandrillus sphinx)", CEFE-CNRS UMR 5175, 1919 Route de Mende, 34293 Montpellier Cedex 5, France, UGENET, CIRMF, Franceville, Gabon, and Department of Biological Anthropology, University of Cambridge, Cambridge, UK, [Behav Ecol 16:614–623 (2005)]

Behavioral Ecology, doi:10.1093/beheco/ari034, Advance Access publication 23 February 2005

[31] Articles cited are: Cockburn A, Osmond HL, Mulder RA, Green DJ, Douvle MC, 2003. Divorce, dispersal and incest avoidance in the cooperatively breeding superb fairy-wren Malurus cyaneus. J Anim Ecol 72:189–202; Griffin AS, Pemberton JM, Brotherton PNM, McIlrath G, Gaynor D, Kansky R, O'Riain J, Clutton-Brock TH, 2003. A genetic analysis of breeding success in the cooperative meerkat (*Suricata suricatta*). Behav Ecol 14:472–480; Mateo JM, 2003. Kin recognition in ground squirrels and other rodents. J Mammal 84:1163–1181; Pusey A, Wolf M, 1996. Inbreeding avoidance in animals. Trends Ecol Evol 11:201–206; Stow AJ, Sunnucks P, 2004. Inbreeding avoidance in Cunningham's skinks (*Egernia cunninghami*) in natural and fragmented habitat. Mol Ecol 13:443–447; Yu XD, Sun RY, Fang JM, 2004. Effect of kinship on social behaviors in Brandt's voles (*Microtus brandti*). J Ethol 22:17–22.

[32] Charpentier, M., Peignot, P., Martine Hossaert-McKey, M., Gimenez,,O., Setchell, J.M. and Wickings, E.J., (2005)

[33] Lukas, D., Reynolds, V., Boesch, C. and Vigilant, L., (2005) Actually the authors claim that this female dispersal is the rule in human groups but I would suggest that this is not an invariable.

[34]Dallwig, Rebecca, "The Common Marmoset, Callithrix jacchus, Current Research." This website reference contains summaries from research described more fully in the following documents: Abbott, D. H., Saltzman, W., Schultz-Darken, N. J., & Tannenbaum, P. L. (1998); Abbott, D. H.,Saltzman, W., Schultz-Darken, N. J., & Smith, T. E. (1997); Baker, J. V., Abbott, D. H., &Saltzman, W. (1999).

[35] [35]Dallwig, Rebecca, "The Common Marmoset, Callithrix jacchus, Current Research.".

[36]Hoier, S.,(2003)

[37] Szwimmer, B. (1995)

"These diagrams are geometric illustrations of incest avoidance prescriptions from Leviticus 18. 'Ego' is the central reference person. Patrilineally related men are in blue; Excluded marriage partners are in red; Partners not explicitly excluded are in green." Szwimmer did not label the members of the clan in his diagram.

[38] Leviticus 18, "Unlawful Sexual Relations",

1 The LORD said to Moses, 2 "Speak to the Israelites and say to them: 'I am the LORD your God. 3 You must not do as they do in Egypt, where you used to live, and you must not do as they do in the land of Canaan, where I am bringing you. Do not follow their practices. 4 You must obey my laws and be careful to follow my decrees. I am the LORD your God. 5 Keep my decrees and laws, for whoever obeys them will live by them. I am the LORD.

6 " 'No one is to approach any close relative to have sexual relations. I am the LORD.

7 " 'Do not dishonor your father by having sexual relations with your mother. She is your mother; do not have relations with her.

8 " 'Do not have sexual relations with your father's wife; that would dishonor your father.

9 " 'Do not have sexual relations with your sister, either your father's daughter or your mother's daughter, whether she was born in the same home or elsewhere.

10 " 'Do not have sexual relations with your son's daughter or your daughter's daughter; that would dishonor you.

11 " 'Do not have sexual relations with the daughter of your father's wife, born to your father; she is your sister.

12 " 'Do not have sexual relations with your father's sister; she is your father's close relative.

13 " 'Do not have sexual relations with your mother's sister, because she is your mother's close relative.

14 " 'Do not dishonor your father's brother by approaching his wife to have sexual relations; she is your aunt.

15 " 'Do not have sexual relations with your daughter-in-law. She is your son's wife; do not have relations with her.

16 " 'Do not have sexual relations with your brother's wife; that would dishonor your brother.

17 " 'Do not have sexual relations with both a woman and her daughter. Do not have sexual relations with either her son's daughter or her daughter's daughter; they are her close relatives. That is wickedness.

18 " 'Do not take your wife's sister as a rival wife and have sexual relations with her while your wife is living."

[39] The same applies to disturbed populations of other species. For instance, in kangaroos, which are frequently and violently 'culled' to 'prevent' kangaroo overpopulation in Australia, e.g. the Belconnen Kangaroo case, where 500 out of 600 kangaroos were killed because it was anticipated that they would overshoot the capacity of the artificially created island they were confined to in a sea of farms and new housing estates in Canberra. The rest were implanted with contraceptives or forcibly sterilized. It was claimed that the kangaroos had tripled their population over the preceding two years. If this were true I surmised that, either the population had grown due to immigration (which had not been considered), or it was an artifact of immigration, and had been artificially formed by dislocation (due to human expansion) of kangaroos all over the state, giving a starting population of highly genetically disparate kangaroos, without the benefit of any Westermarck effect. Had the 600 healthy kangaroos been made up of intact family and clan structures then their fertility would have been curbed by incest avoidance and the Westermarck

60

effect. Unfortunately for these benighted animals, the solution undertaken by the humans who were trying to 'manage' the population, was to drive them randomly into a cul de sac and shoot them. There was no consideration, for humane purposes, and certainly not for Westermarck effect enhancement, of preserving some families intact. And there was no consideration, apparently, that more kangaroos might migrate into the area, since it was not completely impermeable. So, by killing 500 kangaroos indiscriminately, space was liberated for new kangaroo-immigrants from clans all around with no inhibitions about mating with the surviving 100 kangaroos and with each other. By fracturing the clan structures any barriers to fertility that were already present had been removed and space had been created for 500 new and probably unrelated kangaroos. That represents a potential explosion of fertility, even if the contraceptives work. In the mean-time, the kangaroos that suffered the cull have lost the order, protection, security and affection that comes from being part of a clan. Not what you would call humanely or scientifically a win-win solution. "ACT Roo killings: Who profits? Behind the Earless Dragon mask," May 25, (2009). http://candobetter.org/node/1274

[40] This initial explanation does not go into detail of variation in the way that families and societies organise. It was, for instance, quite common for women to have their own villages, where young children of both sexes are raised until the age of about six, when the male children are sent to be raised in the men's villages. This situation may still prevail in some more traditional indiginous societies, especially in remote Pacific Islands or parts of New Guinea.

[41] Abernethy, (2005) pp 64-65.

[42] Levi Strauss on the oppositional structure of incest avoidance, see earlier section.

[43] I first explored this difference between European and Anglophone cultures in my Environmental Sociology thesis, Newman, S.M., The Growth Lobby and its Absence in Australia and France, (2002) notably in Part 2, "Evidence," and in Appendix 3, "Energy and Oil Shocks." I develop the idea more in a book in progress, with the working title of "Lost Tribes, the origins of Capitalism in Britain and of democracy in France," of which this book, The Urge to Disperse, originally formed two chapters. Anyone interested in a nutshell overview of different land-use planning systems might care to look at my work for the Victorian Sustainable Population Australia submission to an Australian Housing Affordability Inquiry, in Newman, S.M., (2003).

[44] This was a term used by Émile Durkheim (1893) to denote societies which were held together 'mechanically' by relatively rigid laws of social conduct and dress, in contrast to 'contractual' societies where laws do not tend to be so

personal, with wide variations in dress and social behaviour, negotiated according to circumstances.

[45] This is the predominant hypothesis explored in B. Diane Chepko-Sade and Zuleyma Tang Halpin, (1987)

[46] Hoier, S. (2003).

[47] Although the Westermarck Effect is never canvassed as such here, a case of a stallion mating with his daughter and of a wolf 'step-father' (who moved in with a female and her cubs after the blood father died) forming a union with the mother's daughter, with the mother leaving the homesite, are described in B. Diane Chepko-Sade, B.D. and Tang Halpin, (1987) p. 50 (in horses) and p.59 (in wolves.) In these cases there had been no opportunity for a Westermarck effect to form.

[48] Macfarlane cites numerous authors on inheritance laws that privileged the first born and allowed disinheritance of the rest. He writes that, by the 13th century, "with regards to freehold land, systemic dispossession of children [had been] incorporated into the law through primogeniture." Alan Macfarlane, *The mystery of Property inheritance and industrialisation in England and Japan*, http://www.alanmacfarlane.com/

The British of the 13th century were struggling against an English legal concept known as Bracton's maxim, *'nemo est heres viventis'* (no one is the heir of a living man), from a case written about by 13th C jurist, Henri de Bracton. British jurist, Maitland, who edited Bracton's Notebook in 1887, commented as follows:

"Free alienation without the heir's consent will come in the wake of primogeniture. These two characteristics **which distinguish our English law from ...[the French]** are closely connected." [My emphasis.] This is a secondary source quote. Henri de Bracton wrote *De legibus et consuetudinibus Angliae* around 1235 but he was still working in 1264 and I do not know the date of the case. Source: Henry de Bracton – Britannica Concise Encyclopedia – wwwa.britannica.com/ebc/article- 9016103, 5/4/2006, 16:27 hrs.

[49] Author unknown, "Aboriginal cultures and the land: An introduction to the unit," p.73. (Accessed 2005)

[50] *Op Cit.*, p.21.

[51] (See also Swimmer, Brian, (1999) on "Parallel Cousin Marriage and Lineage Endogamy" in the notes below.) Such as with brother/sister among Egyptian

62

pharaohs, or in ancient Greek settler societies in African colonies, and among first cousins in Karen [swidden farming, North Thailand] society, to go by tradition elicited in Ferguson, F., (2008), from a Karen folktale.

[52] Swimmer, B. (1995), "The levirate is a widespread institution, which requires that a man becomes the husband of a deceased brother's widow. In the biblical text this imposition is seemingly restricted to a situation in which both brothers reside in the same household and where the deceased has no son to succeed him. It is justified in terms of the need for him to have an heir so that "his name may not be blotted out of Israel(Deuteronomy 25:5)". In this regard, the dead brother rather than the living biological parent becomes the acknowledged or "sociological father" of the child, especially in regard to the establishment of an official genealogical line." (See the story of Judah and Tamara (Genesis 38).

[53] Swimmer, B. (1995) "Parallel Cousin Marriage and Lineage Endogamy

The second substantial prescription is also related to the manipulation of marriage ties in order to ensure continuity within the lineage on occasions in which a man has only daughters. In this case, the daughters inherit his property but are married off to their patrilateral parallel cousins (their father's brother's sons)(Numbers 36). This mechanism allows the property and line of descent to remain within the patrilineage since a daughter's husband belongs to the same lineage as his wife and the children are placed within the patriline through both parents. Accordingly this form of marriage is also referred to as lineage endogamy, i.e., marriage within the lineage.(See Sagas of the Hebrew Patriarchs for a detailed illustration of endogamous lineage development the Hebrew origin myth.)"

[54] Brian Schwimmer, (1995) "Hebrew social organisation : Marriage,"

[55] Farley Mowat, F. (1963) and B. Diane Chepko-Sade and Zuleyma Tang Halpin, (1987).

[56] Hopfenberg and Pimentel, (March 2001)

[57] E.O. Wilson, (2001) "Incest avoidance responses are recognized to occur between people who are raised in close proximity. This has been demonstrated in a 'kibbutz effect' where children raised in the same Israeli kibitzes were not attracted to each other. Several disciplines, prominently including biological anthropology, sociobiology, cognitive psychology, and neuroscience, are yielding evidence that other innate algorithms affect the development of human behavior. These algorithms can be blocked or reversed only at the peril of mental health. An example is the negative imprinting that

forms the basis of incest avoidance, as follows: When either of two persons lives in close domestic proximity during the first 30 months' life of either one, both are unable to form close sexual bonding later in their lives. The phenomenon, known as the Westermarck effect in honor of the Finnish anthropologist who discovered it a century ago, is evidently widespread if not universal in human beings. Equally impressive, it is shared by all other primate species whose sexual behavioral development has been closely studied."